Roadside Guide to Michigan Plants, Trees, and Flowers
An Ecological Approach

Edwin A. Arnfield
Connie Arnfield

Arbutus Press

Dedication

This book is dedicated to our grandchildren: Rachel, Charlotte and Sarah Brumm; Erica, Madeline and Kristen Frogner; Linden McEachern; James Polony, and in loving memory of Brittany Polony. May they continue their active interest in watching over our environment.

Acknowledgements

We wish to thank those who so generously helped us to make this book possible: Pamela Grath, Alan Noftz, James Smith, Debbie Wetherbee, Tom Wetherbee, and Bob Zwemer. Special thanks to our editor, Susan Bays, and to designer Julie Phinney.

Copyright © 2009 Edwin Arthur Arnfield and Constance Joan O'Connell-Arnfield

All rights reserved. No part of this book may be reproduced or transmitted by any means, electronic or mechanical, including photocopying and recording, or by any information storage and retrieval system, except as may be expressly permitted by the 1976 Copyright Act or by the publisher. Requests for permission may be made in writing.

Arbutus Press
www.arbutuspress.com
editor@arbutuspress.com

Book design by Julie Phinney

ISBN 10: 1-933926-12-0
ISBN 13: 978-1-933926-12-4

Table of Contents

Introduction 1
Michigan Geology 2
Glacial Geology 4
Ecological Succession 7
Roadside Plants 9
New Field Stage of Succession 11
    Common Ragweed (*Ambrosia artemisiifolia*) 12
    Lamb's Quarters (*Chenopodium album*) 13
Old Field Stage of Succession 15
    Asparagus (*Asparagus officinalis*) 17
    Big Bluestem (*Andropogon gerardii*) 18
    Black-eyed Susan (*Rudbeckia hirta*) 19
    Chicory (*Cichorium intybus*) 20
    Columbine (*Aquilegia canadensis*) 21
    Common Juniper (*Juniperus communis*) 22
    Dandelion (*Taraxacum officinale*) 23
    Daylily (*Hemerocallis fulva*) 24
    Joe-Pye Weed (*Eupatorium maculatum*) 25
    Milkweed (*Asclepias syriaca*) 26
    Mullein (*Verbascum thapsus*) 27
    New England Aster (*Aster novae-angliae*) 28
    Oxeye Daisy (*Chrysanthemum leucanthemum*) 29
    Plaintain (*Plantago major* and *P. lanceolata*) 30
    Poison Ivy (*Toxicodendron radicans* or *Rhus radicans*) 31
    Queen Anne's Lace (*Daucus carota*) 32
    Salsify (*Tragopogon dubius*) 33
    Spotted Knapweed (*Centaurea maculosa*) 34
    Saint-John's-Wort (*Hypericum perforatum*) 35
    Staghorn Sumac (*Rhus typhina*) 36
    Tall Goldenrod (*Solidago altissima* and *S. canadensis*) 37
    Tickseed (*Coreopsis lanceolata*) 38
Cottonwood Stage of Succession 39
    Bigtooth Aspen (*Populus grandidentata*) 41
    Paper Birch (*Betula papyrifera*) 42
    Quaking Aspen (*Populus tremuloides*) 43
    Red Maple (*Acer rubra*) 44
    Red Oak (*Quercus rubra*) 45
    Serviceberry (*Amelanchier species*) 46
Pine Stage of Succession 47
    Eastern White Pine (*Pinus strobus*) 49
    Jack Pine (*Pinus banksiana*) 50
    Red Pine (*Pinus resinosa*) 51

Oak-Hickory Stage of Succession 53
   Red Oak (*Quercus rubra*) 55
   Shagbark Hickory (*Carya ovata*) 56
   White Oak (*Quercus alba*) 57
Beech-Maple (climax) Stage of Succession 59
   American Beech (*Fagus americana* and *F. grandiflora*) 61
   Basswood (*Tilia americana*) 62
   BloodRoot (*Sanguinaria canadensis*) 63
   Common Violet (*Viola papilionacea*) 64
   Dogwood (*Cornus florida*) 65
   Hop Hornbeam (*Ostrya virginiana*) 66
   Redbud (*Cercis canadensis*) 67
   Sugar Maple (*Acer saccharum*) 68
   Striped Maple (*Acer serotina*) 69
   Trillium (*Trillium grandiflora*) 70
The Transition Forest 71
Marshes and Wetlands 73
   Adder's-Tongue (*Erythronium americanum*) 75
   Balsam Fir (*Abies balsamea*) 76
   Cattails (*Typha latifolia* and *T. angustifolia*) 77
   Eastern Hemlock (*Tsuga canadensis*) 78
   Eastern Tamarack (*Larix laricina*) 79
   Giant Reed (*Phragmites communis*) 80
   Jack-in-the-Pulpit (*Arisaema triphyllum*) 81
   Marsh Marigold (*Caltha palustris*) 82
   Northern White Cedar (*Thuja occidentalis*) 83
   Swamp Milkweed (*Asclepias incarnata*) 84
Beaches and Dunes 85
   Beach Grass/Marram Grass (*Ammophila breviligulata*) 87
   Beach Pea (*Lathyrus japonicus*) 88
   Bearberry or Kinnikinnick (*Arctostaphylos uva-ursi*) 89
   Creeping Juniper (*Juniperus horizontalis*) 90
   Little Bluestem (*A. scoparius/S. scoparium*) 91
   Pitcher's Thistle (*Cirsium pitcheri*) 92
   Prickly Pear Cactus (*Opuntia humifusa* and *O. fragilis*) 93
   Rocket (*Cakile edentula*) 94
   Wormwood (*Artemisia campestris*) 95
Ecology is a Science 97
Index of Latin Names 105
About the Authors 108

INTRODUCTION

This is a field book for those people interested in learning more about the environment in which they live. Specifically, it is for those who travel into Michigan and are curious about the variety of landforms and plants that they see here.

This is not a comprehensive field book, of which there are many available in the market. It is not written for either peers of the author or for experts in the field who know well the detailed scientific sources to be found in relation to this subject.

This book has been written for those of you who travel here for a short time, those who are new arrivals to the area, or perhaps those who have a second home here and would like to be more knowledgeable about the plants in this area. Included are sixty-four interesting plants and trees that you have probably seen many times without knowing their names or any facts about them. That is the reason why it is titled *Roadside Guide to Michigan Plants, Trees, and Flowers: An Ecological Approach*. Most of the plants and trees within this guide can be seen from the roadside or at the beach.

In a guidebook such as this one, there needs to be a unifying theme that aids the user to more easily understand what plants and trees they see here and why they are found in a particular area. Much of what you see will not be found in Florida or some semitropical area, nor can they be seen on prairies or any of the four deserts that are located in the United States. The unifying theme is *succession*.

Understanding primary and secondary succession will allow you to readily identify places where a particular plant or tree can be seen and why it is there in the first place.

Although this book is organized somewhat differently than other guidebooks, it will make it much easier and more interesting for you, no matter where you travel, because the underlying patterns repeat themselves and are easily identified.

MICHIGAN GEOLOGY

In order to better understand the ecology of the Great Lakes area, you need to understand that the history of the earth has affected the area in many ways. Geologic time is divided up into useful segments known as eras, periods, epochs, and lesser units. We understand the earth to be at least four and one half billion years old. The major eras are, from oldest to youngest, the Archeozoic, Proterozoic, Paleozoic, Mesozoic, and Cenozoic. Each is, in turn, subdivided into periods. In many instances certain animals and plants were the dominant form of life during a particular period.

As you can see in the table below, the Devonian Period was dominated by the fishes and the Pennsylvanian by the ferns. But when the glaciers occurred during the Pleistocene, much of that earth rock history was lost by the action of erosion, highlighted in yellow below. The red area in the figure on the next page shows the missing Michigan geological record.

	MICHIGAN BEFORE THE GLACIERS		
Era	Period	Epoch	Time in millions of years ago (mya)
Cenozoic	Quaternary	Recent	begins 4,000 years ago
		Pleistocene	½ to 2 mya **(Ice Ages)**
	Tertiary	Pliocene Miocene Oligocene Eocene Paleocene	**Age of Mammals** begins 63 mya
Mesozoic	Cretaceous Jurassic Triassic	**Age of Reptiles (Dinosaurs)**	
Paleozoic	Permian Pennsylvanian Mississippian Devonian Silurian Ordovician Cambrian	**Age of Amphibians** ends 220 mya Age of Giant Ferns **Age of Fishes**	
Precambrian	Proterozoic/Archeozoic Eras		begins 600 mya

Michigan Geology (continued)

Era	Period	Epoch	Time in millions of years ago (mya)
Cenozoic	Quaternary	Recent	begins 40,000 years ago
		Pleistocene	begins ½ to 2 mya
Unconformity	Missing layers in the Michigan geological record		
Paleozoic	Pennsylvanian		ends 280 mya
	Mississippian		
	Devonian		
	Silurian		
	Ordovician		
	Cambrian		begins 600 mya
Precambrian	Proterozoic and Archeozoic era		4500 mya (origin of the earth)

GLACIAL GEOLOGY

To better understand the history of the earth, geologists divide up geologic time into eras, periods, epochs, and lesser units. We understand the earth to be at least four and a half billion years old. The Pleistocene Epoch—or as some know it, the Ice Age—began about two million years ago. In order for a glacier to develop, temperatures must be cold enough that winter snow accumulation does not melt away in the summer but remains and builds up. The eastern United States, including Michigan, had so much precipitation that it was covered by a vast sheet of ice, amassing as much as ten thousand feet in thickness, which is nearly two miles.

Several glacial and interglacial intervals are known in the Michigan and Great Lakes area. The oldest interval was the Nebraskan, followed by the Kansan, Illinoisan, and finally the Wisconsin. No evidence exists in Michigan for the two oldest glacial periods, but there is evidence south of Michigan in Ohio and Indiana. The Wisconsin glacier covered the region down to nearly the center of the state of Ohio, where the ice margin terminated. The glacier began to melt 125,000 years ago, and the land was completely uncovered about 10,000 years ago. The weight of that ice sheet caused the rock layers beneath it to sink.

Glacial ice causes the land over which it passes to erode and transports vast amounts of rock debris and soil, which are in turn deposited either directly as the glacier melts or indirectly by the movement of the melt waters. As the glacier moves forward, it pushes a mass of rock debris ahead of it,

Glacial Geology (continued)

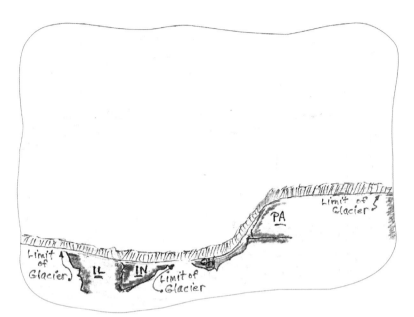

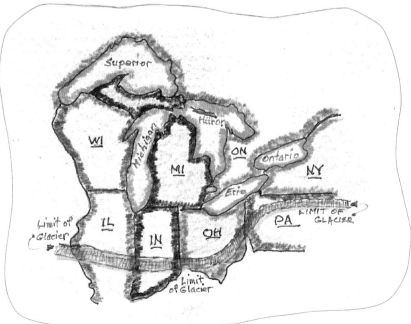

The Last Glacier Recedes
(125,000 Years Ago)

Glacial Geology (continued)

and as it melts back, it leaves debris in the form of terminal moraines and lateral moraines.

To visualize what happened, think of a shovel pushing snow on the sidewalk. If you do not lift the shovel at intervals to toss away the snow, it piles up ahead of the shovel blade and also begins to spill off along the sides. The buildup in front would represent a terminal moraine and along the sides two lateral moraines. There are many other signs of this glacial action, such as pots and kettles and glacial striae gouges on the rock surfaces.

The Great Lakes were also sculpted and formed by glacial action. As drainage altered course from south to east, meltwater ran to the ocean via the St. Lawrence River. The thickness of glacial debris is close to 1,500 feet in the center of Michigan. The huge weight of ice caused underlying soil and rock layers to sink. As the glacier melted, land began to rise slowly, higher in the north than the south of Michigan. This uplifted land affected shore features, such as beaches and wave-cut cliffs. As the land slowly rose, the lakeshores changed, leaving a series of beach and ridge lines along each of the Great Lakes. Much of the soil in northern Michigan is quite sandy because of the debris, giving rise to the types of plants and trees that we see today.

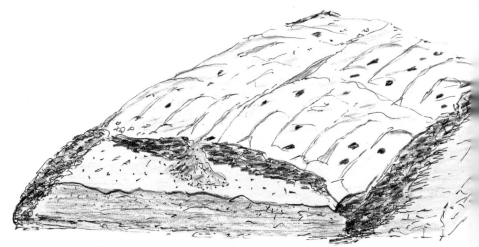

Lateral moraine   Terminal moraine and outwash delta   Lateral moraine

ECOLOGICAL SUCCESSION

"**Science** is a *process* utilizing empirically collected data sets, rationally organized, pragmatically tested and verified through repeatability, whose resulting *product* contributes to the understanding of our universe."

"**Ecology** is the study of living organisms and their interactions with both the *biotic* (living) and *abiotic* (nonliving) environment in which they exist."

Our world is comprised of four spheres: the lithosphere (land masses of the earth), the hydrosphere (the oceans, lakes, rivers, and ground waters), the atmosphere (air mass surrounding the earth), and the biosphere (all the living organisms in, over, and upon its surfaces). Of those four spheres, the biosphere is the most difficult to visualize.

All of the various physical and biological factors influence every organism upon the earth. Those factors determine the variety of ecosystems in which we live. Ecosystems are composed of three groups of organisms: producers, consumers, and decomposers. The producers in the ecosystem are the green plants that convert sunlight into sugar. Consumers are the animals that derive their energy needs by eating energy-rich organisms, such as plants or other animals. Decomposers complete the cycle by releasing the energy and nutrients from dead plants and animals, thus sending nutrients back into the ecosystem. Without this recycling group, the nutrients would remain trapped within plant and animal tissues at their death and would never be reconverted.

We experience change in the daily sequence of night and day and throughout the yearly cycle of seasons. Biologists study the transformation of organisms over great periods of time or the process of evolution through such events as earthquakes and volcanic eruptions. They also study long-term change, such as the weathering of rock into soils or the drifting of continents as they react to alterations in the plates that make up the crust and upper mantle of the lithosphere.

Groups of living organisms in a community change as well, and this process of replacement of species by other species over time is known as *ecological succession*. This succession is an ordered sequence beginning with less stable plant and animal communities that transform gradually until a more stable and permanent climax community evolves. Depending on where you live, the ecological succession in that area follows a predictable pattern.

Ecosystems usually adapt in response to some disturbance that alters their landscape, such as a change in the surface configuration of the ground or in the physical condition of the soil within the system. Ponds fill up with

Ecological Succession *(continued)*

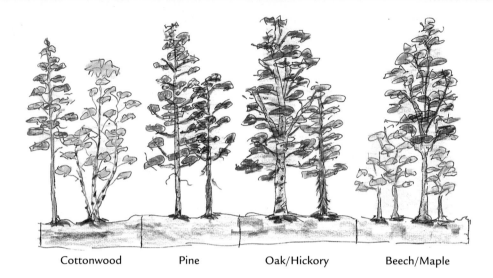

| Cottonwood | Pine | Oak/Hickory | Beech/Maple |

sediment, riverbeds deepen or fill in response to their load of sediment, or soil deposits at the edge of large lakes alter their shape.

Earthquakes, fires, high winds, floods, and epidemics are factors that happen frequently on various parts of the earth, altering the ecosystems in which they occur. In response, the ecosystem undergoes a series of community changes that eventually return the area to conditions prior to the event. The first organisms to invade the disturbed area are plants and animals of the pioneer communities, the last are those of the climax community.

In the midwestern United States, a state such as Michigan can be used to illustrate one pattern of succession. If you were to take the pavement of a large parking lot, bulldoze it to expose the bare soil, and then allow it to recover naturally, a process of ecological succession would begin. If you were able to return to the area 200 or 300 years later, it would have become a climax community of beech and maple trees. In the intervening years, a series of stages would have occurred there, passing from new field to old field, cottonwood, pine, oak-hickory, and finally, beech-maple forest.

Such a radical change, where there is no subsoil of any type, would be called *primary ecological succession*. Normally, in our experience, this does not occur. Instead the system reverts to one of its former stages. That would be known as *secondary ecological succession*, which is most of what we see. A drastic change, such as the eruption of Mount St. Helens, would be an example of a primary succession.

ROADSIDE PLANTS

In the explanation of the earliest stages of succession, the new field stage and the old field stage, you will find that many plants seem to fit into either category of succession. For that reason, a detailed explanation of both stages is given with several examples, and the remaining plants, which seem to fit into either category, are classified for your convenience as roadside plants.

Queen Anne's lace in autumn.

Strawberry Patch

>There is no soul.

But surely there's a hidden inner self
>that no one knows but you.

Mother Nature is a myth.

But there's a rushing sense of something
>in your hands from having worked
>in April's dirt.

And as the raked-out stones were thrown
>onto the brush
>the warmth they'd stolen from the sun
>came into you.

Even with yourself you were at peace.

Though hunger vied with ache to occupy
>your conscious thoughts
>it was suppressed.

The sense that Earth and you are one was stronger still.

That sensing made you smile in silent
>recognition and faded

The smell of coffee took its place.

Ed Arnfield, 1995

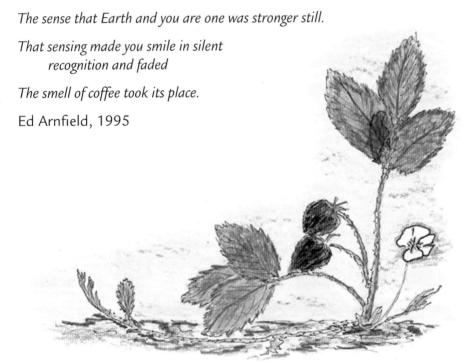

NEW FIELD STAGE OF SUCCESSION

The physical characteristics of the newly formed disturbed environment include constant sunlight, little or no shade, few minerals, and little organic matter in the soil along with extreme dryness.

The first colonizers (pioneer plants) will be those known as *r-selected,* which are species that tolerate such conditions, have a high reproductive rate, and grow rapidly. Such plants include ragweed, lamb's-quarters, goldenrods of various species, and many varieties of annual grasses. They are physiologically hardy species and do not tolerate shade.

This stage may last from one to two years. Many of these plants are easily seen if you examine a pile of overburden (rock or soil overlying a mineral deposit or other underground feature) stripped from the ground at a new building site. By season's end, the pile may be completely covered with such pioneer plants.

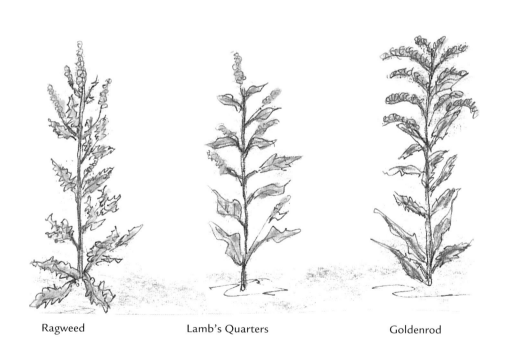

Ragweed   Lamb's Quarters   Goldenrod

Common Ragweed *(Ambrosia artemisiifolia)*

This is one of the plants that most often cause hay fever. The pale green leaves are deeply dissected, and the flowers are green or greenish yellow in color. It often grows to four or five feet in height. Common ragweed is an early invader of disturbed land and fields. Since it is air pollinated, it is a frequent cause of allergic reactions, such as runny eyes and sneezing.

| Mar | Apr | May | June | July | Aug | Sept | Oct |

Roadside Plants — New Field Stage of Succession

Lamb's Quarters *(Chenopodium album)*

Lamb's quarters is a very common roadside weed. This nonnative species came from Europe. It has a diamond-shaped leaf that is green above and whitish underneath, as well as flowers that are green or greenish-yellow in color. For years it was used as a green vegetable in New England, where children were sent out into the field to pick a basketful for dinner. Only the most tender leaves were picked from the plant. It has a high amount of vitamin C. It is a frequent invader of disturbed land, being one of the earliest to arrive.

| Mar | Apr | May | June | July | Aug | Sept | Oct |

Roadside Plants — New Field Stage of Succession

Mullein (left) and milkweed (below) pods in autumn.

OLD FIELD STAGE OF SUCCESSION

A disturbed field will show evidence by the third or fourth year of extensive growth by annuals and perennials. It will begin to support some low and shrubby tree species as well as plants such as mullein, Joe-Pye weed, black-eyed Susans, foxtails, and plantains.

The field will look as though there are scattered, larger plants growing up through the smaller plants and grasses. The amount of shade will slowly increase, which also increases the amount of moisture available. Plants such as annuals that have died in the previous year will have decayed enough to provide organic matter in the soil as well as releasing some accumulated nutrients. Many pioneer plants can no longer grow here because the physical conditions have changed and the ecological succession from new to old field stages has occurred.

Some Old Field Plants

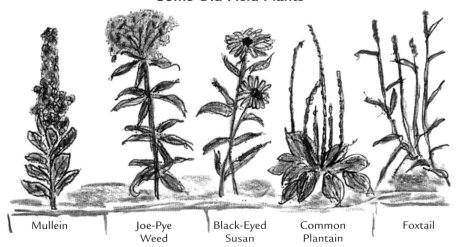

Mullein | Joe-Pye Weed | Black-Eyed Susan | Common Plantain | Foxtail

Some Old Field Shrubs

Staghorn Sumac | Common Juniper

Old Field Stage of Succession

Staghorn sumac in autumn.

Asparagus *(Asparagus officinalis)*

The word asparagus comes from the Greek word asparagos, meaning "to swell." Officinalis is "from the shops," since that is where it is generally found. As a garden crop, it is planted about one foot deep. As the spears appear, the trench is gradually filled in up to ground level. Harvest occurs in the spring for about four weeks. If you continue cutting, the plant will die. It is dioecious, meaning there are male and female plants. In Michigan, people have traditionally gathered wild plants along roadsides in springtime, looking for foliage from last year's plants and cutting the spears of the new growth.

| Mar | Apr | May | June | July | Aug | Sept | Oct |

Roadside Plants — Old Field Stage of Succession   17

Big Bluestem *(Andropogon gerardii)*

This is one of the more beautiful grasses, being tall and stately. Big bluestem is the "type" of grass found on the tallgrass prairie "biome" (climatically and geographically defined area of ecologically similar communities of plants, animals, and soil organisms, also referred to as an ecosystem) in the United States. In Michigan, it is mostly found along highways in the western Lower Peninsula. The stem varies in color from brown or gold to orange or green. This plant blossoms late, usually in September and October. It has fuzzy flower branches that flow out from the top of the stem.

Mar	Apr	May	June	July	Aug	Sept	Oct

Roadside Plants — Old Field Stage of Succession

Black-eyed Susan *(Rudbeckia hirta)*

This is perhaps one of the more familiar garden plants. It is native to Michigan and grows in even the most inhospitable environments. It has a hairy stem and brilliant yellow petals that surround a brown center containing the seeds. Height can vary from two to six feet, depending on soil conditions. The leaves are long, rough, and hairy.

Mar	Apr	May	June	July	Aug	Sept	Oct

Roadside Plants — Old Field Stage of Succession

Chicory *(Cichorium intybus)*

This is a common roadside plant with blue, occasionally pink or white, blossoms. The petals are square-tipped and fringed and are attached directly to the stem. The leaves are dandelion-like close to the ground but smaller on the upper stem. Chicory is often seen as a leafless plant because the upper leaves are so small and the flowers so prominent. It blossoms sequentially, giving the plant a long flowering season. Chicory is prized for its roots, which are a very common additive to coffee in Europe and New Orleans, Louisiana. It has been used as a diuretic and as a laxative.

Mar	Apr	May	June	July	Aug	Sept	Oct

Roadside Plants — Old Field Stage of Succession

Columbine *(Aquilegia canadensis)*

Columbine is a trumpet-like or tube-shaped flower of brilliant red color. It is frequently found along roadsides that have direct sunlight only for part of the day. The flower, which is a drooping bell with spurs, is found at the end of a long stem. The brilliant red color makes it attractive to hummingbirds.

Mar	Apr	May	June	July	Aug	Sept	Oct

Roadside Plants — Old Field Stage of Succession

Common Juniper *(Juniperus communis)*

There are three types of juniper in Michigan. Common juniper is a shrub, creeping juniper is a low-growing type, and upright juniper is much taller and less frequently found here. The needles come in threes and are pointed; mature berries are bluish in color. The flavor of the alcoholic beverage, gin, comes from the essence of these berries. Common juniper grows about four or five feet in height but can spread as wide as ten feet. An area with a heavy density of juniper is nearly impossible to walk through.

| Mar | Apr | May | June | July | Aug | Sept | Oct |

Dandelion *(Taraxacum officinale)*

Dandelion is probably the most recognized roadside flower. It has a yellow flower, jagged leaves, and a very milky and bitter tasting stem. It is a Eurasian native that frequently invades lawns. The seeds are found in a fragile white ball and are dispersed by the wind as "parachutes." The leaves are frequently picked for use as a green vegetable or for making wine.

| Mar | Apr | May | June | July | Aug | Sept | Oct |

Roadside Plants — Old Field Stage of Succession

Daylily *(Hemerocallis fulva)*

Daylily is very commonly found along roadsides during midsummer. This is a plant of Eurasian origin that has escaped from garden cultivation. Reproducing seldom by seed, it usually spreads through underground runners, called rhizomes. The blossom has no spotting, and there is also a yellow variation, often called lemon lily (*H. flava*). Although there can be more than one blossom on a stalk, there is usually only one at a time that lasts only for a day. In French Canada, it is known as *Lis d'un jour*, which freely translates as "lily of only one day." In China, the root extract is said to be an antibacterial agent.

Mar	Apr	May	June	July	Aug	Sept	Oct

Roadside Plants — Old Field Stage of Succession

Joe-Pye Weed *(Eupatorium maculatum)*

Joe-Pye weed is an erect plant growing to four or five feet in height. Its flowers appear yellow or pink and are found on an erect, often purple-spotted stem. The leaves and stem are woolly in character. Native Americans used the infused leaves as an herbal tea for rheumatism and urinary tract infections.

Mar	Apr	May	June	July	Aug	Sept	Oct

Roadside Plants — Old Field Stage of Succession

Milkweed *(Asclepias syriaca)*

This nonnative species is very common along the roadside and in old and new fields. The plant is a favorite of the monarch butterfly (*Danaus plexippus*), whose caterpillars feed on the underside of the leaves before entering their cocoon stage elsewhere.

The juice of the plant is a milky color and gives rise to the plant's name. The seed pods are rather warty in character and open in the late fall, releasing masses of small seeds attached to a filmy upper portion that helps them to disperse in the wind. Many people collect the pods and empty them before spraying them a variety of colors to be used in dry flower displays.

Mar	Apr	May	June	July	Aug	Sept	Oct

Roadside Plants — Old Field Stage of Succession

Mullein *(Verbascum thapsus)*

Common mullein grows to about four or five feet. Its pale to dark green and hairy leaves have a little stem (petiole) and appear attached to the slender stalk. The flowers, which are composed of five petals, bloom progressively from the top downward. This is a nonnative invader from Europe. The infused leaves are used as a tea for asthma, bronchitis, and chest colds.

| Mar | Apr | May | June | July | Aug | Sept | Oct |

Roadside Plants — Old Field Stage of Succession

New England Aster *(Aster novae-angliae)*

This is a very showy flower along roadsides; its violet or purple flower rays are found in clusters at the end of its branches. The flower rays can number as high as eighty or ninety on one blossom. The plant grows three to six feet in height with long lanceolate leaves. Native Americans used the plant to treat fevers.

Mar	Apr	May	June	July	Aug	Sept	Oct

Roadside Plants — Old Field Stage of Succession

Oxeye Daisy (Chrysanthemum leucanthemum)

This is a very familiar white daisy found along roadsides. The flower is white with a yellow center, and the rays are notched at the tips and number as many as twenty-five. There is one flower head on each stem. The narrow leaves have many lobes and are dark in color. The plants grow up to three feet in height and are often found in colonies. This is another nonnative species introduced from Europe.

| Mar | Apr | May | June | July | Aug | Sept | Oct |

Roadside Plants — Old Field Stage of Succession

Plantain *(Plantago major and P. lanceolata)*

Common plantain *(Plantago major)* has its flowers arranged tightly onto a tall stalk, while **English plantain *(P. lanceolata)*** has its flowers at the tip of a cylindrical head at the end of the stalk. The leaves of common plantain are wide and stalked; those of the English plantain are narrow or lance-like. Both are familiar weeds found in lawns. The seeds are sometimes sold in pet stores as canary food. Both plants are nonnative.

| Mar | Apr | May | June | July | Aug | Sept | Oct |

Poison Ivy *(Toxicodendron radicans* or *Rhus radicans)*

The ubiquitous poison ivy plant seems adaptable to numerous habitats but is frequently found along roadsides in the forest margin. The plant is related to the familiar staghorn sumac found in both new and old field habitats. Counting the leaflets is a waste of time since there are so many plants, clovers, raspberries, and strawberries, to name a few, that have the same number of leaves. The berries are whitish in color. The toxic material is the oil of the plant that, in any amount, can cause a reaction in a sensitive person. Avoid it.

| Mar | Apr | May | June | July | Aug | Sept | Oct |

Roadside Plants — Old Field Stage of Succession

Queen Anne's Lace *(Daucus carota)*

Queen Anne's lace is a familiar roadside plant. As its Latin name implies, it is related to the garden carrot. The plant grows two to three feet in height and has a flat, clustered flower of many tiny white florets and usually one purple or black floret at the center. The dried flower curls up into a bird-nest shape and is often used in dried flower arrangements. The leaves are cut into a fern-like shape. The plant is not native; it originated in Europe and was brought to North America by settlers. It has been used in folk medicine as a diuretic.

Mar	Apr	May	June	July	Aug	Sept	Oct

Salsify *(Tragopogon dubius)*

The Latin name comes from the Greek *tragos* (goat) and *pogos* (beard). This plant reaches one to three feet in height and has a yellow flower. When in seed, it appears to be a huge dandelion. The stem is slender as are the leaves, which are grass-like. There is a swelling just below the flower head. Flower bracts are long, pointed, and green in color. It is another nonnative from Europe, where its lower leaves were eaten as greens.

| Mar | Apr | May | June | July | Aug | Sept | Oct |

Roadside Plants — Old Field Stage of Succession

Spotted Knapweed (Centaurea maculosa)

The Latin *maculosa* refers to the spots on the flower head, which is pinkish-purple and resembles a thistle. This plant is also known as wild bachelor's button. It is a wiry plant with thin leaves and grows from one to three feet in height. It is a common sight in fields and is a very noxious invader. This plant is a problem for farmers and ranchers because it exudes a substance from its roots that kills other plants close to it. Interestingly, sheep graze on it and seem to be the only farm animal that does. It is also a nonnative invader.

| Mar | Apr | May | June | July | Aug | Sept | Oct |

Saint-John's-Wort *(Hypericum perforatum)*

This yellow-flowered plant with a cluster of yellow stamens grows to three feet in height. The margins of the petals are spotted with black dots. It has many branches with leaves in pairs, and the five petals are long. The word *wort* is Old English and means plant or root. An infusion of the leaves as tea purportedly is a cure for depression. The plant is nonnative.

| Mar | Apr | May | June | July | Aug | Sept | Oct |

Roadside Plants — Old Field Stage of Succession

Staghorn Sumac *(Rhus typhina)*

Staghorn sumac is a low-growing shrub or small tree with a trunk that is shortened and often crooked. It may grow to twelve or more feet in height. Sumac is intolerant of shade. The flower stalks resemble the horns of a stag, while the leaves are compound, lance-like, and pointed. In autumn they turn a brilliant scarlet and make a showy display anywhere the plant grows in dense groups. The bark is dark brown and hairy. Blossoms are picked when new and then brewed into an herbal tea to treat asthma. It is related to poison ivy (*Rhus radicans*).

| Mar | Apr | May | June | July | Aug | Sept | Oct |

Roadside Plants — Old Field Stage of Succession

Tall Goldenrod (Solidago altissima and S. canadensis)

Goldenrod is often seen in clumps because it reproduces by means of underground roots or rhizomes. There are a great many species of goldenrod in Michigan, so for our purposes it seems easier to classify them as genus *Solidago* rather than by species. The flowers are yellow, the leaves narrow, and most plants are two to three feet high. It is a common roadside plant and easily identified. Look for some stem swellings that are due to gall wasp larvae in the stem. Although it is blamed for causing hay fever, the major culprit is ragweed.

Mar	Apr	May	June	July	Aug	Sept	Oct

Roadside Plants — Old Field Stage of Succession

Tickseed *(Coreopsis lanceolata)*

This has a bright yellow, daisy-like flower with notched petals and each blossom on a separate stem. It grows one to two feet in height. Look for it in large colonies. The lower leaves are long and found on a short stalk.

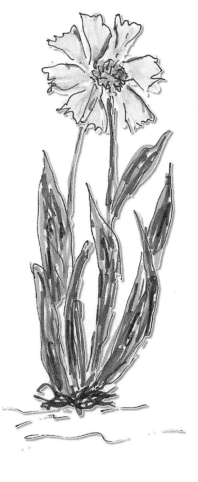

| Mar | Apr | May | June | July | Aug | Sept | Oct |

38   Roadside Plants — Old Field Stage of Succession

COTTONWOOD STAGE OF SUCCESSION

The old field stage of succession may last for as many as twenty or more years but will eventually give way to the cottonwood stage when the first large trees invade the area and begin to spread through the field. Aspens and other cottonwood species, as well as birches and locusts, then appear. As time passes, they increase the amount of shade to a point where fewer and fewer ground cover plants can grow in the absence of intense sunlight. The area undergoes a visual change as well from a rather patchy appearance to one resembling a true woods environment. The soil continues to be enriched and its temperature is cooler. Animals that prefer the new conditions begin to colonize the area. This cottonwood stage will gradually be replaced by a new stage, that of the pines.

The *Populus* species have some unique characteristics for long-term survival. They are capable of establishing new plants in the surrounding areas by means of underground runners or rhizomes. Often there are dozens of younger trees around the older ones, and they are all interconnected. In a drought situation, the groundwater is shared by the entire colony, allowing older trees upslope to survive. In many areas the colonies may be hundreds of years old. One survival advantage is that if fire sweeps through and destroys the colony above ground, the new growth quickly sprouts up from the rhizomes and reestablishes the colony.

Quaking Aspen   White or Paper Birch   Black Locust

Quaking Aspen   Paper Birch   Bigtooth Aspen   Red Maple   Red Oak

Cottonwood Stage of Succession

Bigtooth Aspen *(Populus grandidentata)*

Bigtooth (or large tooth aspen) is a moderately tall tree with a straight trunk and, when growing in a crowded area, has few limbs on its lower branches. It is frequently associated with quaking aspen. The leaves are alternate and short-pointed with large marginal teeth, and its bark is usually green or greenish-grey in color. The trunk can be nearly black at the base of the tree. This is another tree that can reproduce asexually by root suckers or rhizomes. Big tooth aspen is a frequent follower of fires and thrives on poor or sandy soil.

Mar	Apr	May	June	July	Aug	Sept	Oct

Cottonwood Stage of Succession

Paper Birch (Betula papyrifera)

This is a medium-sized tree with alternate and simple leaves that are nearly oval in shape and are whitish or pale on the underside. Its wood was frequently used in the past for thread spools in the textile industry. Paper birch grows tall and thin. When growing in a wooded area, its branches grow close to the top. However, when appearing in more open areas, many branches form on the lower half as well. The outer bark strips easily from the trunk, leaving an orange-colored inner bark. Because the bark can be used in a variety of merchandise, people in northern areas often strip it to sell. This should be discouraged since the bark will never grow back again.

Mar	Apr	May	June	July	Aug	Sept	Oct

Cottonwood Stage of Succession

Quaking Aspen *(Populus tremuloides)*

Quaking aspen is a medium-sized tree with simple alternate leaves that are very finely toothed on the edges and are dark green on top and paler green below. The bark is smooth in young trees and becomes rough, furrowed, and greenish grey in older trees. It is intolerant of heavy shade. The name derives from the fact that in the wind the leaves shimmer or quake rapidly. Much reproduction is by underground roots or rhizomes. A grove of quaking aspen can be hundreds of years old. While fire can destroy mature trees, the extensive root system leads to rapid regrowth within a few years.

Mar	Apr	May	June	July	Aug	Sept	Oct

Cottonwood Stage of Succession

Red Maple *(Acer rubra)*

Red maple lives up to its name, having red flowers in spring as well as red twigs and leaf stems. It is also known as scarlet maple because of the riot of color it displays in autumn. Although it prefers moist soil, it does well in the sandy soils of many areas across northern Michigan. It grows to a height of sixty or more feet and is not a long-lived tree. Although the wood is very similar to that of the sugar maple, it is not an important timber source due to its softness. The bark is reddish-brown in young trees but gradually changes to dark grey as the tree ages.

Mar	Apr	May	June	July	Aug	Sept	Oct

Cottonwood Stage of Succession

Red Oak *(Quercus rubra)*

Red oak is a very widely distributed tree in the eastern United States. Most oak leaves can be divided into two groups depending on whether the ends of the leaf lobes are rounded (as in white oaks) or pointed (as in red oaks). The leaves are alternate and simple, with a dark green color above and more yellow-green below. The horizontal ridgelines seen on the bark are a distinguishing characteristic. It is often found on sites along with aspen and birch since it is only moderately tolerant of shady conditions. Red oak is often used for flooring.

Mar	Apr	May	June	July	Aug	Sept	Oct

Cottonwood Stage of Succession

Serviceberry *(Amelanchier species)*

This is a genus with several species found frequently in Michigan. It has multiple names: serviceberry, juneberry (for when it blooms), and shadbush (because it blooms in many areas at the same time as the fish known as shad run and breed). The leaves are alternate with fine teeth and are oval in shape with pure white blossoms that appear before the tree begins to show its leaves. It is a frequent understory tree in some forests and also occurs regularly on the more open margins of forests. The blossoms have narrow petals and are long. Its fruit turns reddish in the late summer and is sought after by many animals and birds.

Mar	Apr	May	June	July	Aug	Sept	Oct

Cottonwood Stage of Succession

PINE STAGE OF SUCCESSION

Depending on whether the soil is sandy or more dense in nature, several species of pine may establish themselves throughout the cottonwood stage area. The Michigan region supports three species: eastern white pine, red pine, and jack pine (a scrubbier variety than the two taller ones).

    The pines add their leaf litter to the floor of the newly formed forest, and those needles gradually accumulate into a layer of carpet-like litter. Because their needles are acid, they begin to change the soil pH (acidity) to a point where many prior species of trees can no longer invade and compete with the pines. Light is again changed in intensity as the branches of the trees tend to make the ground pattern of light and dark quite patchy. The cones with their seeds begin to provide food for a mixed group of larger animals, many of which take up residence within the tree boughs or in the dense area under low-lying branches. In time, perhaps as long as one hundred years, the debris and soil buildup along with the enlarged populations of residents allow for the development of a new type of forest, that of oak and hickory.

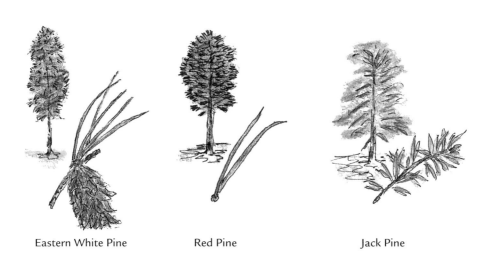

Eastern White Pine   Red Pine   Jack Pine

One aspect of the pine forest litter mass is that the lower pH inhibits the rapid degradation of the humus to release nutrients. If you walk in this forest, you are well aware of the crunchy feel of the needles and the fact that there is little understory vegetation. Slope reforestation with pine species to stop erosion can be problematic. A heavy rainfall can cause runoff to go under the layer of needles and erode the soil to the point that the tree roots have no means to support the tree on the slope. Then, gully erosion clears the slope of the undermined trees.

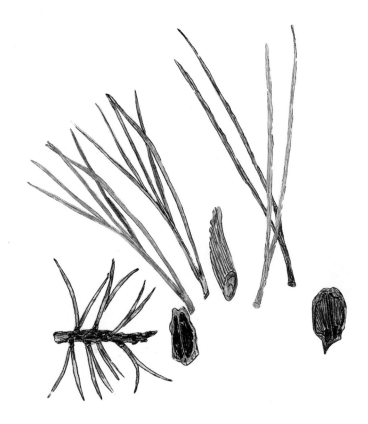

Pine Stage of Succession

Eastern White Pine *(Pinus strobus)*

The eastern white pine is the tallest pine in northern forests. It may reach a height of 170 feet or more. It is easily identified because the bluish-green needles occur in groups of five. It flowers in May and June. The wood is soft and even-textured, making it easy to work, and it is probably the most used wood in house construction. Michigan's lumber industry has been based on the widespread occurrence of this pine. The tops are often bent by the prevailing wind.

Mar	Apr	May	June	July	Aug	Sept	Oct

Pine Stage of Succession

Jack Pine *(Pinus banksiana)*

This is usually a scrubby tree growing on poor soil in Michigan. The needles are arranged in bundles of two and are about two inches long and very stiff. The foliage is a light green and the bark light in color. It is an important habitat for the Kirtland's warbler. The cones are serotinous, meaning they only open after a fire has loosened the sap that holds them closed. Sir Joseph Banks first discovered this pine during an exploration of the new world. He founded the Royal Botanical Gardens at Kew, England, and was president of the Royal Society.

Mar	Apr	May	June	July	Aug	Sept	Oct

Pine Stage of Succession

Red Pine *(Pinus rubra)*

The long needles of this pine occur in groups of two and are dark green in color. The tree is medium-sized and clear of lower branches when grown in a dense habitat. The bark is quite distinctive, being reddish in color and divided into large, scaly plates. The wood is heavier and much stronger than white pine, which makes it a valuable timber tree. It is often referred to as Norway pine but is native to North America. It is a stately tree that can grow to 125 feet in height.

Mar	Apr	May	June	July	Aug	Sept	Oct

Pine Stage of Succession

Place for Thought

The northern hardwood forest in the night
gives refuge to my wish for solitude;
to light a fire and lie where only bright
and ever-distant Autumn stars intrude.

The shifting wind wafts embers toward me now
(they cool before they scorch the greenish cloth)
and dry oak leaves begin to rattle how
they came to meet their recent chilly death.

As floods of patterned thought surge through my head
I sense a draft that causes me to check
my ears—to burrow deeper in my bed
to tuck more padded folds about my neck.

then think again of how a lonely trip
can help me see: one needs companionship.

Ed Arnfield

OAK-HICKORY STAGE OF SUCCESSION

The oak-hickory forest is another step toward the final or climax forest and is often referred to as a pre-climax forest. Such trees as red oak, black oak, and white oak slowly establish themselves beneath the pines of the former stage. In Michigan the hickory is seldom seen above a line drawn roughly from Bay City west to Ludington on the Lower Peninsula. Shade from these trees prevents the growth of new pine seedlings, and they soon overgrow and displace the pine trees that had formerly formed the forest canopy. A vast new array of animals are then established within the oak-hickory forest that feed on debris, the trees themselves, or other animals that take up residence in the forest.

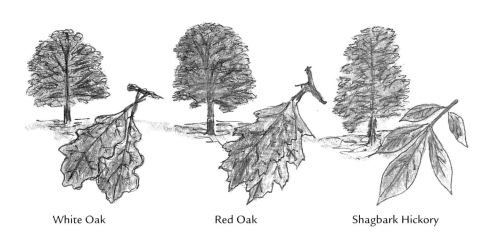

White Oak   Red Oak   Shagbark Hickory

54 Oak-Hickory Stage of Succession

Red Oak *(Quercus rubra)*

Northern red oak is widely distributed in northern Michigan. It reaches a height of seventy to eighty feet but is not tolerant of shade. It is widely used for flooring as the wood is heavy, hard, and very close grained. It belongs to the black oak group because the leaf lobes are pointed instead of round.

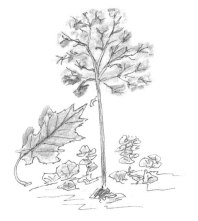

Mar	Apr	May	June	July	Aug	Sept	Oct

Oak-Hickory Stage of Succession

Shagbark Hickory *(Carya ovata)*

The leaves are alternate and compound with five to seven leaflets arranged opposite or sometimes alternately on the leaf stem. The bark is quite distinctive, being light grey in color and arranged in long shaggy plates often loose at both ends. Because of this bark characteristic, it is easily identified at a distance. Any shagbark hickory found in northern Michigan above a line from Bay City west to Ludington in the Lower Peninsula was probably planted by pioneers or settlers in their farmyards.

| Mar | Apr | May | June | July | Aug | Sept | Oct |

Oak-Hickory Stage of Succession

White Oak *(Quercus alba)*

White oak is the most important member of the oak genus since it is the most valuable for commercial purposes. It is easily distinguished from most other oaks because the lobes of the leaves are rounded rather than pointed. It grows on a wide range of sites from moderately sandy to moist bottomlands. The seeds are known as acorns and are a major source of nutrition for a wide range of animals in summer and winter. Because the seed is not bitter, it was used by pioneering families as a source of food and often was ground into flour when other sources were not available.

| Mar | Apr | May | June | July | Aug | Sept | Oct |

Oak-Hickory Stage of Succession

BEECH-MAPLE (CLIMAX) STAGE OF SUCCESSION

Michigan lies within the temperate deciduous forest or eastern deciduous forest biome that occurs in areas of the world that have a climate of warm summers and cool, wet and mild winters. These forests occur extensively in the northern hemisphere, especially North America and are characterized by deciduous species. Ground cover consists of low-growing and frequently early flowering perennials and small shrubs. The understory is mostly young seedlings of taller trees, and there is a canopy of large species that are capable of succeeding themselves. However, as you travel farther north within Michigan, the climax associations become more complex. While southern Michigan has numerous oak-savanna, oak-hickory, and beech-maple communities, farther north the forest is made up of deciduous swamp, pine, conifer bog and swamp, spruce-fir boreal, and northern hardwood communities.

    The dominant tree within most of the eastern deciduous forest is the sugar maple. Farther west the basswoods tend to dominate, while farther south into Arkansas the oaks prevail. Birches, basswood, ash, and elm are frequently included, as well as hop hornbeam and ironwood. The habitat is characterized by mainly cool conditions and usually nutrient-rich soils. Fire is not a frequent problem due to the moist conditions, while within the forest enough fire episodes occur to maintain stands of white pine and other pines over vast areas.

    As stated, there are a variety of climax communities within the eastern deciduous forest. One of those is known as coastal forest. The conditions of water, wind, and sandy soil favor many deciduous species atop the dunes along the shores of the Great Lakes. Here you find a mix of sugar maple, red maple, red oak, birches, and pines. The dune-forming conditions on the eastern shore of Lake Michigan are favorable for this forest type.

American Beech   Sugar Maple

60   Beech-Maple Stage of Succession

American Beech *(Fagus americana* and *F. grandiflora)*

This is a major tree of the eastern deciduous or hardwood forest, and along with the sugar maple, is a dominant tree in this biome. It is a tall tree of eighty or more feet in height. It seems to dominate the forest early but is prone to disease. When it dies and leaves a hole in the upper canopy, a sugar maple usually replaces it. The bark is quite smooth and grey in color and is a favorite for those who carve their initials (often inside the shape of a heart) on trees. Although the wood is heavy and hard, it is most frequently used to make wine and whiskey barrels and is exported to Spain for that purpose.

Mar	Apr	May	June	July	Aug	Sept	Oct

Beech-Maple Stage of Succession

Basswood *(Tilia americana)*

Basswood grows with other hardwoods and is a frequent component of this stage of succession, especially as a dominant tree in northern Michigan, Wisconsin and Minnesota. The heart shaped leaves and the yellowish flowers on top of the long bract make it easy to identify when it blossoms. The wood lacks taste and color and thus was used often for food containers. Because it is one of the softest and lightest of hardwoods, it is valuable for woodcarving and especially for making wooden models in the pattern industry.

| Mar | Apr | May | June | July | Aug | Sept | Oct |

Beech-Maple Stage of Succession

Bloodroot (Sanguinaria canadensis)

Bloodroot is one of the earliest flowers found in the Climax Forest. It is usually in association with adder's tongue in large masses. The rounded leaves are quite large and have a slight blue-green cast to them. The flower petals are white and the center of the blossom is bright yellow. Many are gathered by dyers and weavers because the juice of the root is a bright orange-red. It is much sought after for the deep dye color it yields. Native Americans have ascribed many uses for this plant. It is said to stimulate appetite, lower the temperature of a fever, and helpful to those experiencing bronchitis.

Mar	Apr	May	June	July	Aug	Sept	Oct

Beech-Maple Stage of Succession

Common Violet *(Viola papilionacea)*

There are many variations of this violet as well as numerous garden introduced varieties. Perhaps it is better to refer to them as family *Violaceae*. This plant is found in moist woods as well as other damp places; frequently in the beech-maple association. The leaves are dark green and heart shaped. Although the plant produces seed, it probably reproduces mostly by rhizomes or underground runners.

| Mar | Apr | May | June | July | Aug | Sept | Oct |

Beech-Maple Stage of Succession

Dogwood *(Cornus florida)*

Dogwood is an understory tree in this stage of succession, and it prefers a dry and often hilly area. Such trees thrive where there is at least partial shade during the day. Although Michigan is the far northern edge of its range it can be found in the southern part of the state. This is a prized tree for landscaping due to its spectacular blossoms of white, yellow or pink. These large flower bracts surrounding the greenish flower clusters add to the charm of the blossom. In the northern areas of the state it will only be found in yards and gardens if it is strong enough to survive the extremes of weather.

Mar	Apr	May	June	July	Aug	Sept	Oct

Beech-Maple Stage of Succession

Hop Hornbeam *(Ostrya virginiana)*

Hop hornbeam is frequently found in the understory of this succession, although it can also be found in that of the Oak-Hickory Succession. Because of its hardness the tree is also referred to as Ironwood. The wood is very dense, strong, and close-grained. It is the toughest of native woods but is seldom found in lumberyards because it is only available in small sizes. It is often used for tool handles. The leaves closely resemble those of the birch family of trees. The bark appears deeply furrowed and sometimes looks twisted.

Mar	Apr	May	June	July	Aug	Sept	Oct

Beech-Maple Stage of Succession

Redbud *(Cercis canadensis)*

Redbud is a small tree that prefers moist sites along the banks of streams or on rich bottomlands. It is an understory tree in this succession stage. The pink or red, often called magenta, flowers appear on the branch before the leaves giving the branch a decorated appearance. Michigan is close to the limit of the redbud's northern range, yet the tree is common in the southern parts of the state. It is easily recognized later in the season by the heart shape of the leaves. In the eastern Appalachian mountains redbud appears at lower levels with dogwood but in higher altitudes is replaced by juneberry (*Amelanchier*).

| Mar | Apr | May | June | July | Aug | Sept | Oct |

Beech-Maple Stage of Succession

Sugar Maple *(Acer saccharum)*

The sugar maple, often called hard maple, is probably the most widely distributed tree in the eastern deciduous forest. The leaves are three- or five-lobed with smooth, rounded notches between the sharp ends. They are dark green above and pale on the underside. Young trees have a grayish bark that turns fissured and scaled as the tree matures. This is the source of maple sugar when the sap first runs in the spring. It takes about forty gallons of sap to boil down to a gallon of syrup. This is one of the most valuable commercial hardwoods.

Mar	Apr	May	June	July	Aug	Sept	Oct

Beech-Maple Stage of Succession

Striped Maple *(Acer serotina)*

Striped maple is more shrub-like than tree-like, being an understory species of the hardwood forest. It is easily distinguished by its three-lobed, shallowly notched, and pointed leaves. In addition, the bark has striping that alternates between green and white. For this reason, some call it the Michigan State University tree since those are the university colors. In the far north, it is sometimes known as "moose maple" because moose are attracted to its leaves in the summer. The wood has little commercial value since it is rather soft in texture.

Mar	Apr	May	June	July	Aug	Sept	Oct

Beech-Maple Stage of Succession

Trillium *(Trillium grandiflora)*

Probably the largest flower in its group, it is pure white and turns a pale pink as it ages. Some that are infected by an organism or a virus produce a central green stripe in the flower. The leaves are in threes, and the petals are in threes as well. It grows well in alkaline or neutral acidity soils. Sometimes in spring, it can carpet the ground beneath a beech-maple forest in a spectacular display of green and white.

Mar	Apr	May	June	July	Aug	Sept	Oct

Beech-Maple Stage of Succession

THE TRANSITION FOREST

The eastern deciduous forest gives way over time to a transition forest of approximately 150 miles in width. This forest is a gradual transition from deciduous forest to a true boreal forest of the far north. This transition forest begins as you travel north of a line between Bay City on the east and Ludington on the west of the Lower Peninsula.

In this transition, deciduous types gradually give way to a forest of various spruces, balsam fir (*Abies balsamea*) and eastern hemlock (*Tsuga canadensis*) found in New England, northern New York, Wisconsin, Minnesota, and the upper half of Michigan. In autumn this transition forest is a riot of reds, yellows, oranges, golds, and browns, which are accentuated when displayed against the green-black color of the hemlocks. People from all over the world travel to these areas for "color tours" since no other forest on earth exhibits such an autumnal display.

On the northern margin, the forest gradually becomes the true Canadian evergreen forest (taiga, boreal forest). The spruces, hemlocks, and firs grow thickly, excluding many deciduous species. This can be a cold, dark, and often gloomy habitat, but the evergreens have useful adaptations for this climate. Their branches emerge from the bole or trunk of the tree at right angles, so that when the boughs are snow-laden and heavy, the limbs drop and cause much of the snow to shed or fall off the branches. In addition, the leaves are thin, waxy, and needle-like. This protects them from the snow and ice that forms on the branches. Although we refer to them as "evergreen," they shed some needles all year long, and much of their needles in the early fall. You can see and feel the mat of needles when walking through this type of forest.

Deciduous trees, such as beeches or maples, have a different emergent pattern of branches, many of which angle upward (dendritic pattern). When snow or ice accumulates on them, the weight often causes them to break at the trunk, which makes them unsuitable for this type of climate.

Hammerin' Bird

*The Downy and Hairy
I'm sure find it scary.*

*Not the din that he makes
but the effort it takes
just to hammer and pound
rotted trunks on the ground
as square holes he does make
the odd insect to take.*

*Pileated cousin
the woodlands arousin'
have you shock-absorbed head
so your brain won't go dead?*

Ed Arnfield, 2006

MARSHES AND WETLANDS

Autumn cattails.

Adder's-Tongue *(Erythronium americanum)*

Adder's-tongue is easily identified by the mottled leaves in shades of green and red. After blooming the plants wither above ground so that later in the summer they cannot be seen. Other names for the plant are trout lily and dogtooth violet (although it is not really a violet). They are found in rich deciduous woods and are frequently found in beech-maple associations in early spring. This is a member of the lily family, and flower parts are to be found in threes and sixes. Native Americans used the leaves as a tea for fevers and to prevent conception.

Mar	Apr	May	June	July	Aug	Sept	Oct

Marshes and Wetlands

Balsam Fir *(Abies balsamea)*

Balsam fir is a relatively short-lived tree found in wetlands. The bark exhibits conspicuous blisters on a pale grey-green background. It has a shallow root system and is prone to uprooting by strong winds. The leaves are needle-like and flat, with a rounded tip, and they look similar to hemlock but are not stalked like the hemlock's. The wood is light and soft with a pale yellow color. It is used for lumber and also as pulp for the paper industry. Balsam fir is a favorite Christmas tree because it holds its needles well after cutting.

Mar	Apr	May	June	July	Aug	Sept	Oct

Marshes and Wetlands

Cattails *(Typha latifolia* and *T. angustifolia)*

The two species of cattail in northern Michigan are the common and the narrow-leafed. These are familiar plants found in wet roadside ditches and marshy areas. They have male and female flowers with no petals or sepals and are massed together into long, sausage-like clusters on the upright stems. In the common cattail, the male and female flowers appear in series without a gap between them on the stem. There is a distinct gap between them in the narrow-leafed species. It is a good way to separate the species when you see them. Native Americans used the spreading roots or rhizomes as a source of flour.

Mar	Apr	May	June	July	Aug	Sept	Oct

Marshes and Wetlands

Eastern Hemlock *(Tsuga canadensis)*

This is a tree indicative of the transition forest of northern Michigan. It reaches sixty to seventy feet in height and has many branches when grown in a more open environment. When hemlock stands are dense, the forest floor is very dark since little light enters from above due to the widely branching limbs of the tree. The needles are short with two white lines on the underside. The leaf is stalked; no other eastern evergreen has such a leaf. Hemlock is used primarily for pallets, boxes, and crates.

Mar	Apr	May	June	July	Aug	Sept	Oct

Marshes and Wetlands

Eastern Tamarack *(Larix laricina)*

Eastern tamarack (also called American larch) is a medium sized tree with a height of about sixty-five feet. The interesting characteristic of this "evergreen" is that it sheds all of its leaves or needles in autumn, which is rare. The needles are light green and arranged in bunch-like clusters. Its primary use economically is for posts, railway ties, telephone poles, and any outdoor purpose where resistance to rot is needed.

Mar	Apr	May	June	July	Aug	Sept	Oct

Giant Reed *(Phragmites communis)*

Giant reed is a tall grass, sometimes reaching a height of ten to twelve feet. It is found in and around wetlands and moist areas of waste ground. Although native to North America, it is a very invasive plant in Michigan. It reproduces by means of stolons (over-the-ground runners) and rhizomes (underground runners), which have been measured in lengths of up to seventeen feet. This plant seems to thrive in areas where humans have disrupted or cleared land. It frequently competes with cattails and can lead to their rapid disappearance in wetlands. It is known as an excellent cellulose source for a variety of products made of paper.

| Mar | Apr | May | June | July | Aug | Sept | Oct |

Marshes and Wetlands

Jack-in-the-Pulpit *(Arisaema triphyllum)*

Jack-in-the-pulpit has an upright, curved hood covering a yellow club or spathe, which is covered by large leaves. It has three leaflets on a long stalk. It is usually found in damp woods or marshes. Native Americans used the root as a treatment for colds and coughs. The leaves contain calcium oxalate, which causes an intense burning if eaten raw. Many leaves used for pot herbs contain this substance.

Mar	Apr	May	June	July	Aug	Sept	Oct

Marshes and Wetlands

Marsh Marigold (Caltha palustris)

Marsh marigold, also called cowslip, is characterized by large, glossy, heart-shaped leaves with hollow stems and brilliant yellow flowers, usually with five petals. The flowers of this plant resemble those of a large buttercup. They often grow in large colonies and can cover a considerable area along wet ground or a roadside drain. They are seldom found as individual plants. Early colonists made a tea sweetened with maple syrup to treat coughs.

| Mar | Apr | May | June | July | Aug | Sept | Oct |

Marshes and Wetlands

Northern White Cedar *(Thuja occidentalis)*

The northern white cedar is included in a group known as *Arborvitae* (tree of life). The tea made from the needles or leaves was used by Native Americans as a cure for scurvy (lack of vitamin C), hence the name tree of life. The leaves are scaly, pointed, and flat, while the twigs are arranged in flattened sprays. The wood is one of the most resistant to decay and thus of use for poles, posts, and shingles. It was frequently used in boat building, especially for canoes. It thrives on limestone soils and is oftentimes found in wet or swampy areas.

Mar	Apr	May	June	July	Aug	Sept	Oct

Marshes and Wetlands

Swamp Milkweed *(Asclepias incarnata)*

This three- to four-foot plant is found in wet areas, and its dark red or pinkish flowers are held in an umbrella-like cluster. Just as its name indicates, the sap is milky and bitter to taste (tasting not recommended). It is a favorite of the monarch butterfly and is one of the preferred plants it uses to deposit its eggs. The larvae feed from the underside of the leaf and are seldom seen until they are large and multicolored. The seed pods are long and contain silky seeds that easily float in the wind. Early colonists used the potentially toxic plant to treat asthmatic conditions.

Mar	Apr	May	June	July	Aug	Sept	Oct

BEACHES AND DUNES

In reality, beaches are deserts. The underlying material of a beach is not true soil but a loose aggregation of sand, pebbles, and fragmented rocks. As a result the covering of plants is rather sparse. Dune areas are found immediately behind the beach area and are partially protected from the higher storm waves. They are a product of the sand from the beach being carried by wind off the lakes.

A beach can be separated into fore, middle, and back. The fore beach is constantly under attack by wind and water and constantly changing. This area is characterized by wet sand or gravel and an absence of plants. At the top of the fore beach is a wrack line, a zone of debris consisting of plant and animal materials deposited there by the higher wave action. This wrack area changes according to the wave pressure from day to day.

The middle beach is of a similar nature but is unaffected by normal wave action, although storm waves do have an effect upon the sparse plant life occurring here in low density. There is usually a low dune with a cover of a variety of plants, such as rocket, Pitcher's thistle, little bluestem, Indian grass, creeping juniper, bearberry, and others. Often there is a trough behind this area that is lower in elevation yet contains the same plants.

At an even higher area, the back dune is where plants such as taller shrubs and larger trees begin to establish themselves. Here you will find birches, aspens, red oak, and red maple. Often this area will repeat itself again as you reach the ridgeline.

This sequence of areas is a series of succession stages from bare ground to established climax or coastal forest.

Pussy willow.

Beach Grass/Marram Grass *(Ammophila breviligulata)*

This is a common grass found on most beaches around the Great Lakes. As a matter of fact, it is a common grass found along the ocean shores of the world. The leaves are found atop a long stalk. The color of this body varies from pale yellow to brown as the seeds mature. One of the fascinations of this plant is the series of concentric circles made in the sand by the leaf tips as the wind blows. Because of its extensive root system, it is a very important plant for stabilizing the Great Lakes dunes.

Mar	Apr	May	June	July	Aug	Sept	Oct

Beaches and Dunes

Beach Pea *(Lathyrus japonicus)*

This common beach and dune plant is a low creeper that reaches about two feet in length and is characterized by long clusters of pale pink, violet, or white pea-like blossoms. The tip of the long stalk usually has a tendril that allows it to attach vine-like to other plants for support. Its leaves appear to be attached directly to the plant although they do have a very short petiole or stalk. The blossoms will remind you of snapdragons or sweet peas.

| Mar | Apr | May | June | July | Aug | Sept | Oct |

Beaches and Dunes

Bearberry or Kinnikinnick *(Arctostaphylos uva-ursi)*

Bearberry is a low growing and creeping evergreen and is found frequently providing ground cover on beaches and dunes. The white to pinkish blossoms appear as clusters at the ends of branches. The leaves are thick, shiny, and leathery. In late autumn or early winter, the red berries stand out dramatically against the yellow of the sand underneath. Kinnikinnick was a type of tobacco smoked in peace pipes by many of the eastern tribes of Native Americans.

| Mar | Apr | May | June | July | Aug | Sept | Oct |

Beaches and Dunes

Creeping Juniper *(Juniperus horizontalis)*

Creeping juniper is a very common plant on fore dunes, back dunes, and open sandy areas close to the ridgeline. It has a low creeping, mat-like structure and contains many radiating branches. The branchlets are irregularly arranged on the larger branches and contain sharp, pointed needles at the end in overlapping pairs. When growing closely together, they can form an almost endless mat of green.

Mar	Apr	May	June	July	Aug	Sept	Oct

Beaches and Dunes

Little Bluestem *(A. scoparius/S. scoparium)*

Little bluestem (*Andropogon scoparius/Schizachyrium scoparium*) is a "type plant" of the mixed grass prairie yet is also found extensively along the shores of Lake Michigan. It is a beautiful midsize grass with a many-colored group of stems. Early in growth it is bluish in color, but as it matures, the colors range from tan to brown to dark red. It is usually found in isolated clumps, however, given more room it can become a sod grass. It is an excellent forage grass. The single seed heads are very light and fluffy in appearance.

| Mar | Apr | May | June | July | Aug | Sept | Oct |

Beaches and Dunes

Pitcher's Thistle *(Cirsium pitcheri)*

Pitcher's thistle is listed as a threatened species in Michigan. It should not be disturbed in its habitat, which consists of fore dune, back dune, and the beach areas between them. It is an erect plant standing three feet in height with pale white to pinkish blossoms. The plant lives four to eight years before blossoming, after which it dies. The leaves are branched and covered with spines as well as a grey-green color that is easily distinguished from other plants in the wormwood family. It is well adapted to its harsh and windy environment.

| Mar | Apr | May | June | July | Aug | Sept | Oct |

Prickly Pear Cactus *(Opuntia humifusa/O. fragilis)*

Few people would expect to see native cactus in Michigan, however prickly pear is native to the coastal dunes of the Great Lakes, especially those of Lake Michigan. It has been found in sandy fields, on plains, and sometimes on sunny rocky surfaces. These photographs are of a windblown fragment I saw one autumn over nineteen years ago. I picked it up and since it was not in its native area on the beach or dunes, I replanted it, not thinking it would survive. These cacti are quite adaptable to a number of habitats other than those to which they are native.

| Mar | Apr | May | June | July | Aug | Sept | Oct |

Beaches and Dunes

Rocket *(Cakile edentula)*

Rocket is a member of the mustard family and is a succulent plant of half a foot to one foot in height. It contains small whitish or pale purple flowers. When the blossoms fall there is left a white, hollow ball-like pod that appears similar to the nose cone of a rocket, thus the plant name.

Mar	Apr	May	June	July	Aug	Sept	Oct

Wormwood *(Artemisia campestris)*

Wormwood is a tall, one- to two-foot high plant somewhat similar to ragweed in appearance. The leaves are tiny, narrow, rather stringy and forked, with a grayish green color. The flowers are green to yellow and clustered along the stem. Plants of the wormwood family are toxic in nature and in France were made into a yellowish alcoholic beverage known as absinthe.

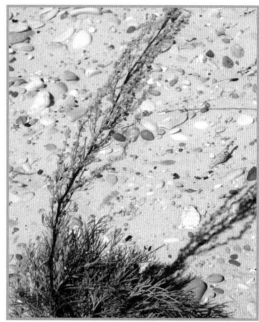

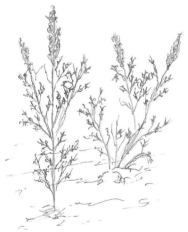

| Mar | Apr | May | June | July | Aug | Sept | Oct |

Beaches and Dunes

The Point

*This wind-whipped saucer of a lake has many faces
seldom quite the same.*

*Oft I walk where Point comes down to meet the lake.
Here eastern waves come crashing in
as though the concrete blocks were each and every one
personal enemies
to be demolished by wind-sent blows of water
five feet high
and crested by the breeze:*

*Driving,
Relentless,
Ceaseless.*

*The land comes to an end.
It disappears
and ever-scouring waves can meet.*

*A place where one can walk
until there's no place left to go.
Land's End.*

*The waves crest
then crash
eating up the sandy spit in shallow channels.
No more they ebb
they rush across.
The spray begins to coat my leg with pain
the surging water hurts.
Now my feet no longer sense the stones.*

*No one here
just gulls
shrill cries
wings against wind
high waves
misting noise
soaking breeze
and me.*

*I walk down the Point
It is no more
I stand within the lake.*

ECOLOGY IS A SCIENCE

Every plant described within this book is a part of an ecosystem. As such, an understanding of some basic principles of ecology and ecosystem succession can help you to better recognize both the plant you are looking at and how and why it is an integral part of that system. The following discussion will help you to better see why science is such an important part of our everyday lives, because whether we like it or not, or understand it or not, for a great many years now our lives have been governed and directed by the outcomes and discoveries of this discipline. The word science comes from an earlier Latin word, *scientia*, which means to know or to understand. So let us then define science as a sort of basic curiosity about everything within us and around us (our environment). As human beings, we have been accumulating vast amounts of knowledge and information about that environment for over two million years. However, the most dramatic expressions of that scientific knowledge have come about within the past several hundred years, and most of that within the last one hundred years.

The definition of science shown above is only one of many that have accumulated over a great number of years. Here are some others:

> "Science is ordered knowledge of natural phenomena and the relations between them."
>
> William C. Dampier, 19th Century

> "Science is the process which makes knowledge.
>
> Charles Singer, 19th Century.

> "Science is the search for the reason of things."
>
> Havelock Ellis, 20th Century.

> "Science is trained and organized common sense."
>
> Thomas H. Huxley, 19th Century.

> "Science is knowledge not of things but of their relations."
>
> Henri Poincare, 19th Century.

> "Natural science is a way of describing reality; it is therefore limited by the limits of observation; and it asserts nothing which is outside observation. Anything else is not science, it is scholastics."
>
> Jacob Bronowski, 20th Century.

And finally,

> "Science is a process utilizing empirically collected data sets, rationally organized, pragmatically tested, verified through repeatability, whose resulting product contributes to the understanding of our universe."
> Edwin A. Arnfield, 21st Century.

Humans are a species that likes order. So, if we attempt to categorize those definitions by grouping together those that are similar in nature, they fall within two groupings. The first group defines science as a body of knowledge, a group of concepts, or a set of ideas; the second group defines science as a means of gaining new knowledge. In other words, science seems to be comprised of both product and process. The product is that gigantic accumulation of knowledge, information, hypotheses and theories that appear to explain how our environment works, while the process lays out a systematic approach to how we expand that awareness.

We have gathered such a vast amount of knowledge that we tend to subdivide and categorize it into groups of related ideas and knowledge that we call disciplines. Biology, the study of living organisms of all types as well as their functional mechanisms, is one of those disciplines. Within it we have a subgroup known as ecology. We will look at a subdivision of ecology which is known as ecological succession.

There are other disciplines such as chemistry, physics, geology, and astronomy, as well as a multitude not mentioned here. Each of those disciplines have also been divided, giving us subdivisions of biology such as ecology, anatomy, physiology, microbiology, vertebrate and invertebrate biology, and even smaller groupings such as entomology, protozoology, and virology.

Now we need to look at the second part, the process, because science teaches us problem-solving methods that we can use throughout our lives. Most scientists would call that the process of inquiry. Another name for it is the scientific method, which is usually subdivided into a series of five steps:

1. An awareness and definition of a problem.
2. The process of data gathering, accumulation and categorization.
3. The generation of a hypothesis, or tentative and possible problem solution.
4. A substantial period of controlled testing and experimentation.
5. The generation of a theory from the supported hypothesis.

Most problem-solving situations commence when the experimenter begins to ponder seriously some question about phenomena within a particular field of work or discipline. This can be a long and frustrating period until the experimenter is able to narrow the scope of the problem to one or several carefully constituted questions. From that point on, the experimenter collects data, facts, and information pertaining to the question and places them into some usable order system. Ultimately the experimenter uses sense data accumulated by careful observation. The ordering of the data usually culminates in the generation of a good educated guess as to the solution. That guess is refined and formed into what is called a hypothesis. That statement is the basis for testing the validity of the tentative solution to the problem. Then testing or experimentation follows. The experimenter attempts to exclude all other possible solutions to the problem by stating the problem in a way that leads to only one pair of conclusions, regardless of their validity. This tentative solution or hypothesis is or is not supported by the evidence.

The hypothesis is now called a theory. If supported, the theory will explain the problem in such a way that a solution has been found. However, the theory is valid only so long as no new data come to light that are unexplainable by that theory. If new data disagrees with, or can not be explained by the theory, then a new solution and a new theory must be generated. One must explain the phenomena and does so by including the new data.

The methods of science are both inductive and deductive. The inductive process proceeds by working from the parts to the whole; it is the generation of an idea that explains a group of disparate or separate phenomena. After such an induction process has generated a theory, a process of deduction can be utilized. The deductive process proceeds by working from the whole to the parts and is used to explain each instance where the theory explains a similar phenomenon. You may know of induction as synthetic reasoning and deduction as analytic reasoning.

If science can be used to explain various natural phenomena, then it can be viewed as a philosophy. Philosophy is a way of viewing and explaining how the world works. It is also a basic set of guidelines for living one's life. Each of us between the ages of sixteen to the early twenties evolves some group of principles upon which to base our lives. Evolving such principles seems to be a very necessary step in each adolescent human's existence. If we can show that science and its associated methods can help to explain natural phenomena in the world around us, then we can make a case that science is indeed a philosophy. Or, if we can show that science is comprised of other valid philosophies, we can prove that science is a philosophy. The first approach has already been explained, the second approach needs some more thought.

Several philosophies have interesting points when viewed in relation to science. Empiricism is a philosophy that views the universe in terms of the human senses. If you use those senses, you should be capable of explaining natural phenomena. Francis Bacon, a seventeenth-century empiricist, is remembered for his emphasis upon the inductive process of reasoning. His *New Organon* was written as a treatise against Aristotle's logic of scientific inquiry, the *Organon*. Bacon sensed that in seeking answers to the "fixed laws" of physical phenomena, for example the nature of heat, you first made a careful collection and arrangement of all the situations in which heat occurs. The resulting table or list followed the idea that "the cause of anything is present when the observed effect is present." Bacon then would list a second grouping that included all the instances where heat was absent, and a third grouping where there were variations in the amount of heat. By comparison of the three tables, incorrect explanations would be rejected. The final step was a process of elimination.

John Locke, an eighteenth-century philosopher and probably the foremost English empiricist, wrote that "No man's knowledge can go beyond his experience." Our outward knowledge of the world is perceived through our senses, and he believed that an understanding of our environment could be accomplished because of that sense perception.

Children play many games as they grow up. One they frequently engage in at some time in their lives is what I would label "What if you lost this sense?" We seem to be aware that our outward world is analyzed in terms of our ability to see, hear, taste, smell, and touch that world. The game choices are very difficult since your ultimate survival depends upon your senses functioning in continuous group interaction. How would you function in an environment where you could not see, hear, feel, taste, or smell? Knowledge does begin with sensation. Philosophically however, not all aspects of your environment can be explained by those senses.

Rationalism is another philosophy of interest to us. It is a word derived from the Latin word, *ratione*, "to reason." This philosophy holds that you can understand your environment in terms of the ability of the human mind to reason. The human brain is capable of remarkable activity in terms of turning sense data into a comprehensible world. Rene Descartes, a seventeenth-century French philosopher, was a brilliant advocate of such a philosophy. The various activities of inductive and deductive reasoning, also known as synthetic and analytic reasoning, are derivations of human reasoning capabilities. Descartes defined rationalism as a philosophy in which all knowledge is deduced from one or a few fundamental concepts, usually without recourse to experience. The rationalist would rely most heavily upon the deductive processes, and proposes that the human mind is capable of perceiving truth without the necessity of testing the results. One criticism of empirical science was that it left no room for judgment. As biologist George G. Simpson states the problem:

> "In reality, gathering facts, without a pretty good idea as to what the facts mean, is a sterile occupation and has not been the method of any important scientific advance. Indeed, facts are elusive and you usually have to know what you are looking for before you can find one."

Yet reasoning alone cannot wholly describe our environment.

A third and final philosophy is pragmatism. This philosophy holds that you can understand the way the world works in terms of consequences. Americans are a very pragmatic group. They pride themselves as "practical." Does it work? Good! If it does not work, abandon it. Pragmatism looks at the world in terms of cause and effect. If I do this, that will happen. Charles Sanders Peirce, an American twentieth-century philosopher, saw pragmatism as "a method for making our ideas clear." Peirce insisted, as Aristotle did before him, "that we know things by their sensible effects. A thing is called hard because it is not easily scratched; it is called heavy because it will fall." William James, another pragmatist, defines "true" as "only the expedient in our way of thinking, and "right" as "only the expedient in our way of behaving." The pragmatist John Dewey listed five steps to reflective thinking: a felt difficulty, its location and definition, suggestion of a possible solution, development by reasoning of the bearings of the suggestion, and further observation and experimentation leading to its acceptance or rejection. This pragmatic approach to problem solving can work quite successfully. Yet it does not hold the key to the solution of every problem. It can be misused, as many despotic rulers have, in which the end justifies the means.

All three of these philosophies appear to be a part of the scientific method when you look at the various steps. Empiricism underlies the awareness of a problem and the data gathering activities. Rationalism is a part of defining the problem, generating the hypothesis, and constructing a theory. Pragmatism is an integral aspect of any testing and experimentation. Science possesses components of several philosophies and stands by itself as a means to interpret the environment in a valid manner. The process called the scientific method is a practical problem-solving mechanism. By using it, you can become a more effective solver of everyday problems.

If you think science is capable of solving all problems, you are mistaken. We need to ask the questions, "What is the domain of science?" What does it include? What can science do or not do for us? Some say that science can deal with any phenomenon to which the methods of science can be applied now or in the future. Thus, science has limitations in terms of what it can do.

First, science cannot deal with absolutes or with absolute truths, for if they are already "known," how can science help you to know them?

Second, science cannot make value judgments or moral decisions. Beauty, love, hate, virtue, justice, existence, and truth are human values and phenomena that science cannot define. Many of these ideas are discussed and evaluated by an area of philosophy known as metaphysics. Classic metaphysics deals with the highly abstract, the subtle, the abstruse and in its extremes deals with the imaginary and the fanciful. The latter extreme of metaphysics appeals to those who reject science as a problem solving mechanism and look for explanations outside the realm of the testable. Yet, the philosopher Alfred North Whitehead defines such extreme metaphysics as "an attempt to explain the incredible by an appeal to the unintelligible."

Third, science is limited to those phenomena which repeat or are repeatable. It cannot effectively examine onetime events.

Lastly, science is limited by the observer. We speak frequently about objectivity and subjectivity while convincing people that their own feelings and emotions should be excluded as much as possible from scientific investigation. No matter how hard we try, our feelings become a part of what we do. Science, after all, is done by humans. In addition, error occurs in work. We all make errors.

The inherent honesty of the scientific endeavor aids us in that discoveries are published freely as they occur. All others can then see not only what was learned, but more importantly, how it was learned. This openness is a self-policing aspect of current scientific research endeavors. It helps to make science and the pursuit of science an inherently honest exercise.

Another point that must be made if you wish to understand science concerns our expectations of science to solve our problems. The frustrations generated by expectations unrealized have led to an anti-science movement that is quite vocal in its denunciation of the role of science in our modern society.

The movie *Jurassic Park* was the cinematic re-creation of animals that have been extinct for sixty-five million years. It enchanted viewers, making the movie a potent vehicle for the anti-science message: "Look at this terrible thing scientists have done. They have created monsters." Then there was the movie *Lorenzo's Oil*. Its theme was that if science cannot help you to cure your child then you need to search for other unscientific means. Dr. Arthur Caplan, Director of the Center for Bioethics at the University of Minnesota, cited three myths perpetuated by such movies:

> "Our culture wants desperately to believe that cures can be found if only bureaucratic red tape is gotten out of the way. The movie reflects the strongly held American belief that perseverance, hard work and love can conquer any ailment. Finally there is the theme that mainstream science is indifferent and uncaring of the suffering of patients and their families."

To understand science is to realize that it is a powerful tool for solving problems, yet is not the "magic wand" that can solve every problem presented. The limitations of science ensure that scientists cannot go beyond the data before them, representing what can be seen and measured about the problem that they are examining. Science, although a universal tool for understanding, is not a universal remedy for our every problem.

With all these limitations it is fair to ask, "What can science do?" The job of science, according to Philip Kitcher, "is not to hold up an ideal of intellectual purity, but simply to provide an efficient means of uncovering validity in nature." It is not a cookbook sequence either. A scientist and a cook both learn from past experiences how hard a four minute egg will be each time it is cooked, but the scientist is not content to stop there. The scientist wants to know "how" it occurs. From such questions and careful study have come ideas and concepts that are our basis for knowledge about proteins, their properties and their internal molecular structure. Unlike the cook, the scientist can explain what is happening to a four-minute egg and from that experience predict how other similar proteins will act under like circumstances. Furthermore, the scientist may be able to produce new and different molecules of protein.

All of this can be traced back to medieval philosophy. For example, in one instance the explanation for the movement of a thrown missile or stone was that it continues to move because the person who threw it has given the stone the quality of *virtus derelicta* (Latin: force left behind, i.e., the stone retains a propellent force from the thrower). This is a concept similar to our present day notion of momentum and impetus; but merely giving names to motion does not help us to understand the phenomenon. The medieval philosopher's quest ends where it began. Modern physics only begins when we break that circle. Momentum is defined in terms of mass times velocity. We discover how it is acquired, how it is conserved, and how it is dissipated through the actions of measurement, observation and experimentation.

Had the concept of momentum been invalid scientists would search for another concept that was correct. The important shift here is from the question of why the stone moves, which has no answer, to the question of how it moves, a phenomenon immediately observable.

Science is both product and process; it is both a body of knowledge, concepts and ideas and a process for acquiring new knowledge. It is limited in its domain by one-time events, absolutes, moral decisions, value judgments, repeatability and human fallibility. Within these limitations it is a powerful tool responsible for steady human progress toward the ultimate understanding of our surrounding biological and physical environment.

INDEX OF LATIN NAMES

A. scoparius (little bluestem) 91
Abies balsamea (balsam fir) 76
Acer rubrum (red maple) 44
Acer saccharum (sugar maple) 68
Acer serotina (striped maple) 69
Amelanchier species (serviceberry) 46
Ambrosia artemisiifolia (common ragweed) 12
Ammophila breviligulata (beach grass/marram grass) 87
Andropogon gerardii (big bluestem) 18
Aquilegia canadensis (columbine) 21
Arctostaphylos uva-ursi (bearberry or kinnikinnick) 89
Arisaema triphyllum (jack-in-the-pulpit) 81
Artemisia campestris (wormwood) 95
Asclepias incarnata (swamp milkweed) 84
Asclepias syriaca (milkweed) 26
Asparagus officinalis (asparagus) 17
Aster novae-angliae (New England aster) 28
Betula papyrifera (paper birch) 42
Cakile edentula (rocket) 94
Calthra palustris (marsh marigold) 82
Carya ovata (shagbark hickory) 56
Centaurea maculosa (spotted knapweed) 34
Cercis canadensis (redbud) 67
Chenopodium album (lamb's quarters) 13
Chrysanthemum leucanthemum (oxeye daisy) 29
Cichorium intybus (chicory) 20
Cirsium pitcheri (Pitcher's thistle) 92
Coreopsis lanceolata (tickseed) 38
Cornus florida (dogwood) 65
Daucus carota (Queen Anne's lace) 32
Erythronium americanum (adder's tongue) 75
Eupatorium maculatum (joe-pye weed) 25
F. grandiflora (American beech) 61
Fagus americana (American beech) 61
Hemerocallis fulva (daylily) 24
Hypericum perforatum (Saint-John's-wort) 35
Juniperus communis (common juniper) 22
Juniperus horizontalis (creeping juniper) 90
Larix laricina (eastern tamarack) 79

Lathyrus japonicus (beach pea)	88
Opuntia humifusa/O. fragilis (prickly pear cactus)	93
Ostrya virginiana (hop hornbeam)	66
P. lanceolata (plaintain)	30
Phragmites communis (giant reed)	80
Pinus banksiana (jack pine)	50
Pinus rubra (red pine)	51
Pinus strobus (eastern white pine)	49
Plantago major (plantain)	30
Populus grandidentata (bigtooth aspen)	41
Populus tremuloides (quaking aspen)	43
Quercus alba (white oak)	57
Quercus rubra (red oak)	45
Quercus rubra (red oak)	55
Rhus typhina (staghorn sumac)	36
Rudbeckia hirta (black-eyed susan)	19
S. canadensis (tall goldenrod)	37
S. scoparium (little bluestem)	91
Sanguinaria canadensis (bloodroot)	63
Solidago altissima (tall goldenrod)	37
T. angustifolia (cattails)	77
Taraxacum officinale (dandelion)	23
Thuja occidentalis (northern white cedar)	83
Tilia americana (basswood)	62
Toxicodendron radicans or *Rhus radicans* (poison ivy)	31
Tragopogon dubius (salsify)	33
Trillium grandiflora (trillium)	70
Tsuga canadensis (eastern hemlock)	78
Typha latifolia (cattails)	77
Verbascum thapsus (mullein)	27
Viola papilionacea (common violet)	64

Edwin Arthur Arnfield

Ed is a retired college and university professor. For thirty-two years he taught biology at Macomb Community College in Warren and Sterling Heights, Michigan, as well as at Wayne State University and Oakland University as an adjunct faculty member. Prior to that time, he was a research associate in the medical school of the Ohio State University. He has taught every grade but kindergarten in several Michigan public school districts.

He is a graduate of the chemical and biological curriculum at Cass Technical High School, Detroit, Michigan. He holds a B.A. degree in biology and an M.S. in biology from Wayne State University in Detroit, Michigan, and a Ph.D. from Ohio State University, Columbus, Ohio. He occasionally works as a consulting field ecologist in Leelanau County, Michigan, as well as a docent for the Leelanau Conservancy.

He is a life member of the Appalachian Trail Conservancy.

Constance J. O'Connell-Arnfield

Connie is a retired clinical psychologist who worked in private practice for twenty-three years. She is a graduate of Cooley High School, Detroit, Michigan. She graduated from Macomb County Community College with an associate of arts degree and holds a bachelor of arts degree and master of arts degree from Oakland University in Rochester, Michigan. Since retirement she has become an avid photographer, quilter, and gardener.

Both are Leelanau County residents.